YOUR KNOWLEDGE HAS VALUE

- We will publish your bachelor's and
  master's thesis, essays and papers

- Your own eBook and book -
  sold worldwide in all relevant shops

- Earn money with each sale

Upload your text at www.GRIN.com
and publish for free

Jagadish Kumar Mogaraju

Practical Manual for RS and GIS with Quantum GIS

GRIN Publishing

Bibliographic information published by the German National Library:

The German National Library lists this publication in the National Bibliography; detailed bibliographic data are available on the Internet at http://dnb.dnb.de .

Imprint:

Copyright © 2013 GRIN Verlag, Open Publishing GmbH
Print and binding: Books on Demand GmbH, Norderstedt Germany
ISBN: 978-3-656-57880-2

This book at GRIN:

http://www.grin.com/en/e-book/267283/practical-manual-for-rs-and-gis-with-quantum-gis

GRIN - Your knowledge has value

Since its foundation in 1998, GRIN has specialized in publishing academic texts by students, college teachers and other academics as e-book and printed book. The website www.grin.com is an ideal platform for presenting term papers, final papers, scientific essays, dissertations and specialist books.

Visit us on the internet:

http://www.grin.com/

http://www.facebook.com/grincom

http://www.twitter.com/grin_com

YSR ENGINEERING COLLEGE OF YOGI VEMANA UNIVERSITY

PRACTICAL MANUAL FOR RS AND GIS

Compiled by **MOGARAJU JAGADISH KUMAR** M.Tech, Academic Consultant, Department of Civil Engineering, YSREC of YVU, Proddatur, INDIA

M JAGADISH KUMAR

[2012-13]

1. GEOREFERENCING OF TOPOSHEET

Aim: To georeference the given toposheet by using georeferencer.

Introduction:

Georeferencing is the process of assigning real-world coordinates to each pixel of the raster. These coordinates are obtained by doing field surveys - collecting coordinates with a GPS device for few easily identifiable features in the image or map. In case of digitize scanned maps, the coordinates can be obtained from the markings on the map image itself. Using these sample coordinates or GCPs (Ground Control Points), the image is warped and made to fit within the chosen coordinate system.

Methodology:-

- Click on start button on the desktop.
- Select programs and go to Quantum GIS Lisboa.
- Then go to Quantum GIS Desktop 1.8.
- Click on raster on menu bar and go to georeferencer and again click on georeferencer.
- Change the transformation settings.
- Select appropriate transformation type.
- Select appropriate resampling method.
- Select target SRS to give coordinate reference system.
- Add toposheet by clicking on open raster button.
- Click on add points button and add ground control points in Degrees Minutes and Seconds mode at extreme corner points in pixel zoom mode by using pan, zoom in, zoom out buttons.
- The points can be edited in the GCP table which can be opened through view tab.
- Now click on start georeferencing button.
- Select the name and destination to save the file.
- Click on ok button and check the box load into canvas when finished.(Automatically the georeferenced toposheet is loaded into the canvas)
- Close the georeferencer window.
- The georeferenced toposheet will be stored in TIFF format.

- To view the georeferencered toposheet, now go to add raster button and click on to zoom to layer extent.
- Compare the co-ordinates with the toposheet.
- If the coordinates are correct then the toposheet is said to be georeferenced.

Result:

The toposheet is georeferenced as per the coordinates marked with required transformation and hence can be used for further analysis.

2. DIGITIZATION OF TOPOSHEETS USING POINT FEATURE

Aim: To digitize the given toposheet using point feature

Introduction:

- The process of representing an analogue signal or an image by a discrete set of its points is known as Digitizing. This data after conversion is in the binary format, which is directly readable by computer. The analogue signals are variable whereas the digital format is the discrete one. These discrete units are called as bits. These bits organized in groups are known as byte. The digital signals are mainly represented in the form of sequence of integers. These integers can be converted back to analogue signal that are approximately similar to the original analogue signals. Digitizing is done by reading an analogue signal at regular time intervals, representing the value of that point by an integer. The process of digitization is classified into manual digitizing, heads up digitizing, interactive tracing method and automatic digitizing. Manual Digitizing is done using digitizing tablet. The operator manually traces all the lines from his hardcopy map and creates identical digital map on the computer. It is very time consuming and level of accuracy is also not very good. Heads-up Digitizing is similar to manual digitizing in the way that lines have to be drawn manually but directly on the computer screen. So in this level of accuracy increases and time taken decreases. Interactive tracing method is improvement over Heads-up digitizing in terms of speed and accuracy. Automatic Digitizing is automated raster to vector conversion using image processing and pattern recognition techniques In this technique computer traces all the lines, which results in high speed and accuracy along with improved quality of images.

Methodology:-

- Click on start button on the desktop.
- Select programs and go to Quantum GIS Lisboa.
- Then go to Quantum GIS Desktop 1.8.
- Click on add raster layer on tool bar
- Add georeferenced toposheet which is saved as .tif file.
- Right click on the layer in TOC and opt for zoom to layer extent.
- Go to layer on menu bar and click on new.
- Select new shape file layer.

- Then select point type.
- Fill the details in the new vector layer window and select name of the file, type, width to be provided.
- Coordinate reference system should be selected and it should be same as the georeferenced toposheet.
- Click on ok.
- Select the destination for saving the file.
- Make right click on the project name and open the attribute table.
- Select toggle editing.
- Then click on add columns.
- Fill the details like name of the column, type, width to be provided.
- Click on ok and close the window.
- Go to toggle editing and click on add features.
- Select point features and provide ID and name.
- Save the file by clicking on save project.
- Click on file on menu bar and go to new print composer.
- Go to layout and click on add image to add an image, add label to add title, add legend to add a legend, add scale bar to add a scale bar and add North arrow via add image tool.
- Go to file and click on export as pdf for saving the file in pdf format or click on export as image to save the file as image.
- Provide the name for the output file and select the destination for saving the file.

Result:

The point features like spot heights and villages were digitized using the point features.

3. DIGITIZATION OF TOPOSHEETS USING LINE FEATURE

Aim: To digitize the given toposheet using line feature

Introduction:

- The process of representing an analogue signal or an image by a discrete set of its points is known as Digitizing. This data after conversion is in the binary format, which is directly readable by computer. The analogue signals are variable whereas the digital format is the discrete one. These discrete units are called as bits. These bits organized in groups are known as byte. The digital signals are mainly represented in the form of sequence of integers. These integers can be converted back to analogue signal that are approximately similar to the original analogue signals. Digitizing is done by reading an analogue signal at regular time intervals, representing the value of that point by an integer. The process of digitization is classified into manual digitizing, heads up digitizing, interactive tracing method and automatic digitizing. Manual Digitizing is done using digitizing tablet. The operator manually traces all the lines from his hardcopy map and creates identical digital map on the computer. It is very time consuming and level of accuracy is also not very good. Heads-up Digitizing is similar to manual digitizing in the way that lines have to be drawn manually but directly on the computer screen. So in this level of accuracy increases and time taken decreases. Interactive tracing method is improvement over Heads-up digitizing in terms of speed and accuracy. Automatic Digitizing is automated raster to vector conversion using image processing and pattern recognition techniques In this technique computer traces all the lines, which results in high speed and accuracy along with improved quality of images.

Methodology:-

- Click on start button on the desktop.
- Select programs and go to Quantum GIS Lisboa.
- Then go to Quantum GIS Desktop 1.8.
- Click on add raster layer on tool bar
- Add georeferenced toposheet which is saved as .tif file.
- Right click on the layer in TOC and opt for zoom to layer extent.
- Go to layer on menu bar and click on new.
- Select new shape file layer.

- Then select line type.
- Fill the details in the new vector layer window and select name of the file, type, width to be provided.
- Coordinate reference system should be selected and it should be same as the georeferenced toposheet.
- Click on ok.
- Select the destination for saving the file.
- Make right click on the project name and open the attribute table.
- Select toggle editing.
- Then click on add columns.
- Fill the details like name of the column, type, width to be provided.
- Click on ok and close the window.
- Go to toggle editing and click on add features.
- Select line features and provide ID and name.
- Save the file by clicking on save project.
- Click on file on menu bar and go to new print composer.
- Go to layout and click on add image to add an image, add label to add title, add legend to add a legend, add scale bar to add a scale bar and add North arrow via add image tool.
- Go to file and click on export as pdf for saving the file in pdf format or click on export as image to save the file as image.
- Provide the name for the output file and select the destination for saving the file.

Result:

The line features like roads, railways and contours were digitized using the line feature.

4. DIGITIZATION OF TOPOSHEETS USING POLYGON FEATURE

Aim: To digitize the given toposheet using polygon feature.

Introduction:

- The process of representing an analogue signal or an image by a discrete set of its points is known as Digitizing. This data after conversion is in the binary format, which is directly readable by computer. The analogue signals are variable whereas the digital format is the discrete one. These discrete units are called as bits. These bits organized in groups are known as byte. The digital signals are mainly represented in the form of sequence of integers. These integers can be converted back to analogue signal that are approximately similar to the original analogue signals. Digitizing is done by reading an analogue signal at regular time intervals, representing the value of that point by an integer. The process of digitization is classified into manual digitizing, heads up digitizing, interactive tracing method and automatic digitizing. Manual Digitizing is done using digitizing tablet. The operator manually traces all the lines from his hardcopy map and creates identical digital map on the computer. It is very time consuming and level of accuracy is also not very good. Heads-up Digitizing is similar to manual digitizing in the way that lines have to be drawn manually but directly on the computer screen. So in this level of accuracy increases and time taken decreases. Interactive tracing method is improvement over Heads-up digitizing in terms of speed and accuracy. Automatic Digitizing is automated raster to vector conversion using image processing and pattern recognition techniques In this technique computer traces all the lines, which results in high speed and accuracy along with improved quality of images.

Methodology:-

- Click on start button on the desktop.
- Select programs and go to Quantum GIS Lisboa.
- Then go to Quantum GIS Desktop 1.8.
- Click on add raster layer on tool bar
- Add georeferenced toposheet which is saved as .tif file.
- Right click on the layer in TOC and opt for zoom to layer extent.
- Go to layer on menu bar and click on new.
- Select new shape file layer.

- Then select polygon type.
- Fill the details in the new vector layer window and select name of the file, type, width to be provided.
- Coordinate reference system should be selected and it should be same as the georeferenced toposheet.
- Click on ok.
- Select the destination for saving the file.
- Make right click on the project name and open the attribute table.
- Select toggle editing.
- Then click on add columns.
- Fill the details like name of the column, type, width to be provided.
- Click on ok and close the window.
- Go to toggle editing and click on add features.
- Select polygon features and provide ID and name.
- Save the file by clicking on save project.
- Click on file on menu bar and go to new print composer.
- Go to layout and click on add image to add an image, add label to add title, add legend to add a legend, add scale bar to add a scale bar and add North arrow via add image tool.
- Go to file and click on export as pdf for saving the file in pdf format or click on export as image to save the file as image.
- Provide the name for the output file and select the destination for saving the file.

Result:

The area features like forest, water body, barren lands and contours were digitized using the polygon feature.

5. CREATION OF DIGITAL ELEVATION MODEL

Aim: To create Digital Elevation Model using point feature.

Introduction:

Digital Elevation Models arc data files that contain the elevation of the terrain over a specified area, usually at a fixed grid interval over the bare Earth. The intervals between each of the grid points will always be referenced to a specific geographical coordinate system. This is either latitude-longitude or UTM (Universal Transverse Mercator) coordinate systems. The closer together the grid points are located, the more detailed the information will be in the file. The details of the peaks and valleys in the terrain will be better modeled with small grid spacing than when the grid intervals are very large. Several factors are important for quality of DEM-derived products terrain roughness, sampling density (elevation data collection method), Grid resolution or pixel size, Interpolation algorithm, Vertical resolution and Terrain analysis algorithm. The uses of DEMs include extracting terrain parameters, modeling water flow or mass movement, creation of relief maps, rendering of 3D visualizations, creation of physical models, rectification of aerial photography or satellite imagery, reduction (terrain correction) of gravity measurements and terrain analyses in geomorphology and physical geography

Methodology:-

- Click on start button on the desktop.
- Select programs and go to Quantum GIS Lisboa.
- Then go to Quantum GIS Desktop 1.8.
- Click on add raster layer on tool bar and add a modified tif file.
- Open toposheet and digitize it.
- Go to layer on menu bar and click on new.
- Select new shape file layer.
- Then select point type.
- Fill the details in the attribute table like name of the file, type, width to be provided.
- Click on ok.
- Select the destination for saving the file.
- Make right click on the project name and open the attribute table.
- Select toggle editing.
- Then click on add columns.
- Fill the details like name of the column, type, width to be provided.
- Click on ok and close the window.

- Go to toggle editing and click on add features.
- Select spot heights and provide ID and name.
- Save the file by clicking on save project.
- Go to plugins on menu bar and click on it.
- Then go to manage plugins and select interpolation plugin.
- Click on interpolation icon on tool bar.
- Give vector layers and attribute ID.
- Open interpolation plugin. Add vector layer and click on add button.
- Select interpolation attribute and Interpolation method.
- Select output file (TIN) and click on ok.
- Click on raster.
- Go to analysis and then select DEM (Terrain models).
- Open DEM and select output file.
- Select appropriate mode and vertical exaggeration.
- Click on ok.
- Make right click on DEM and zoom to layer extent.
- Then Digital Elevation Model (DEM) will be created.
- If you want to change the colour of DEM, make right click on DEM.
- Then go to properties and change color map as pseudocolor in single band properties.
- Click on ok.
- Then the required Digital Elevation Model will be created.
- Save the file by clicking on save project.
- Click on file on menu bar and go to new print composer.
- Go to layout and click on add image to add an image, add label to add title, add legend to add a legend, add scale bar to add a scale bar and add North arrow via add image tool.
- Go to file and click on export as pdf for saving the file in pdf format or click on export as image to save the file as image.

- Provide the name for the output file and select the destination for saving the file.

Result:

The Digital Elevation Model is created using spot heights as point features and can be used for further analysis.

6. CREATION OF CONTOURS THROUGH DIGITAL ELEVATION MODEL

Aim: To create contours by using Digital Elevation Model.

Introduction:

Digital Elevation Models are data files that contain the elevation of the terrain over a specified area, usually at a fixed grid interval over the bare Earth. The intervals between each of the grid points will always be referenced to a specific geographical coordinate system. This is either latitude-longitude or UTM (Universal Transverse Mercator) coordinate systems. The closer together the grid points are located, the more detailed the information will be in the file. The details of the peaks and valleys in the terrain will be better modeled with small grid spacing than when the grid intervals are very large. Several factors are important for quality of DEM-derived products terrain roughness, sampling density (elevation data collection method), Grid resolution or pixel size, Interpolation algorithm, Vertical resolution and Terrain analysis algorithm. The uses of DEMs include extracting terrain parameters, modeling water flow or mass movement, creation of relief maps, rendering of 3D visualizations, creation of physical models, rectification of aerial photography or satellite imagery, reduction (terrain correction) of gravity measurements and terrain analyses in geomorphology and physical geography. The contours with a specified interval can be generated using DEM.

Methodology:-

- Click on start button on the desktop.
- Select programs and go to Quantum GIS Lisboa.
- Then go to Quantum GIS Desktop 1.8.
- Click on add raster layer on tool bar and add DEM layer from the saved location.
- Click on raster on menu bar.
- Then go to extraction and select contour and provide sufficient contour interval.
- Provide output file name and select the destination for saving the file and click on ok.
- Right click on the layer and go to properties.
- Then click on style and select colormap.
- Open colormap, enter the number of entries and click on classify.
- Then click on apply.
- Right click on the layer and click on zoom to layer extent.
- Save the file by clicking on save project.

- Click on file on menu bar and go to new print composer.
- Go to layout and click on add image to add an image, add label to add title, add legend to add a legend, add scale bar to add a scale bar and add North arrow via add image tool.
- Go to file and click on export as pdf for saving the file in pdf format or click on export as image to save the file as image.
- Provide the name for the output file and select the destination for saving the file.

Result:

The contours were created with specific intervals using DEM.

7. DETERMINATION OF HILLSHADE FOR THE GIVEN TERRAIN

Aim: To determine hillshade by using Digital Elevation Model for the given terrain.

Introduction:

The Hillshade function obtains the hypothetical illumination of a surface by determining illumination values for each cell in a raster. It does this by setting a position for a hypothetical light source and calculating the illumination values of each cell in relation to neighboring cells. It can greatly enhance the visualization of a surface for analysis or graphical display, especially when using transparency. To calculate the shade value, the altitude and azimuth of the illumination source is needed. The altitude of the illumination source is specified in degrees above horizontal. The direction of the illumination source, azimuth, is specified in degrees.

Methodology:-

- Click on start button on the desktop.
- Select programs and go to Quantum GIS Lisboa.
- Then go to Quantum GIS Desktop 1.8.
- Click on add raster layer on tool bar and add DEM layer from the saved location.
- Click on raster on menu bar.
- Then go to terrain analysis and select hillshade.
- Provide output file name and select the destination for saving the file and click on ok.
- Right click on the layer and go to properties.
- Then click on style and select color map.
- Open color map, enter the number of entries and click on classify.
- Then click on apply.
- Right click on the layer and click on zoom to layer extent.
- Save the file by clicking on save project.
- Click on file on menu bar and go to new print composer.
- Go to layout and click on add image to add an image, add label to add title, add legend to add a legend, add scale bar to add a scale bar and add North arrow via add image tool.

- Go to file and click on export as pdf for saving the file in pdf format or click on export as image to save the file as image.
- Provide the name for the output file and select the destination for saving the file.

Result:

The Hillshade effect of the given area is determined by using Digital Elevation Model.

8. DETERMINATION OF SLOPE FOR GIVEN TERRAIN

Aim: To determine slope by using Digital Elevation Mode for the given terrain.

Introduction:

Slope identifies the steepest downhill slope for a location on a surface. The lower the slope value, the flatter the terrain; the higher the slope value, the steeper the terrain. The slope output raster map contains slope values, stated in degrees of inclination from the horizontal if format equal to degrees option (the default) is chosen, and in percent rise if format is equal to percent option is chosen. Slope is the rate of maximum change in z-value from each cell. The use of a z-factor is essential for correct slope calculations when the surface z units are expressed in units different from the ground x,y units. The range of slope values in degrees is 0 to 90. For percent rise, the range is 0 for near infinity. A flat surface is 0 percent, a 45 degree surface is 100 percent, and as the surface becomes more vertical, the percent rise becomes increasingly larger.

Methodology:-

- Click on start button on the desktop.
- Select programs and go to Quantum GIS Lisboa.
- Then go to Quantum GIS Desktop 1.8.
- Click on add raster layer on tool bar and add DEM layer from the saved location.
- Click on raster on menu bar.
- Then go to terrain analysis and select slope.
- Provide output file name and select the destination for saving the file and click on ok.
- Right click on the layer and go to properties.
- Then click on style and select colormap.
- Open colormap, enter the number of entries and click on classify.
- Then click on apply.
- Right click on the layer and click on zoom to layer extent.
- Save the file by clicking on save project.
- Click on file on menu bar and go to new print composer.

- Go to layout and click on add image to add an image, add label to add title, add legend to add a legend, add scale bar to add a scale bar and add North arrow via add image tool.
- Go to file and click on export as pdf for saving the file in pdf format or click on export as image to save the file as image.
- Provide the name for the output file and select the destination for saving the file.

Result:

The slope of the area is determined by using Digital Elevation Model of the given terrain.

9. DETERMINATION OF ASPECT FOR GIVEN TERRAIN

Aim: To determine aspect by using Digital Elevation Mode for the given terrain.

Introduction:

Aspect identifies the down slope direction of the maximum rate of change in value from each cell to its neighbors. Aspect can be thought of as the slope direction or the compass direction a hill faces. The values of the output raster will be the compass direction of the aspect. Aspect is the direction of the maximum rate of change in the z-value from each cell in a raster surface. Aspect is expressed in positive degrees from 0 to 359.9, measured clockwise from north.

Methodology:-

- Click on start button on the desktop.
- Select programs and go to Quantum GIS Lisboa.
- Then go to Quantum GIS Desktop 1.8.
- Click on add raster layer on tool bar and add DEM layer from the saved location.
- Click on raster on menu bar.
- Then go to terrain analysis and select aspect.
- Provide output file name and select the destination for saving the file and click on ok.
- Right click on the layer and go to properties.
- Then click on style and select colormap.
- Open colormap, enter the number of entries and click on classify.
- Then click on apply.
- Right click on the layer and click on zoom to layer extent.
- Save the file by clicking on save project.
- Click on file on menu bar and go to new print composer.
- Go to layout and click on add image to add an image, add label to add title, add legend to add a legend, add scale bar to add a scale bar and add North arrow via add image tool.
- Go to file and click on export as pdf for saving the file in pdf format or click on export as image to save the file as image.

- Provide the name for the output file and select the destination for saving the file.

Result:

The aspect is determined by using Digital Elevation Model of the given terrain.

10. DETERMINATION OF RELIEF FOR GIVEN TERRAIN

Aim: To determine relief by using DEM of the given terrain.

Introduction: Relief can be used to calculate the amount of sun or shade for a 3D surface. This analysis uses a fixed location of the sun and the horizon to accurately display areas of bright sun exposure as well as low dark areas that contain lots of shadow. Typically a shaded relief will be used in presentation of 3D GIS analyses as a thematic background layer that provides the user with pretty looking cartographic representation. A relief result provides the most visually appealing display of the 3D DEM data.

Methodology:-

- Click on start button on the desktop.
- Select programs and go to Quantum GIS Lisboa.
- Then go to Quantum GIS Desktop 1.8.
- Click on add raster layer on tool bar and add DEM layer from the saved location.
- Click on raster on menu bar.
- Then go to terrain analysis and select relief.
- Provide output file name and select the destination for saving the file and click on ok.
- Right click on the layer and go to properties.
- Then click on style and select colormap.
- Open colormap, enter the number of entries and click on classify.
- Then click on apply.
- Right click on the layer and click on zoom to layer extent.
- Save the file by clicking on save project.
- Click on file on menu bar and go to new print composer.
- Go to layout and click on add image to add an image, add label to add title, add legend to add a legend, add scale bar to add a scale bar and add North arrow via add image tool.
- Go to file and click on export as pdf for saving the file in pdf format or click on export as image to save the file as image.

- Provide the name for the output file and select the destination for saving the file.

Result:

The relief is determined by using Digital Elevation Model of the given terrain.

Aim: To determine ruggedness by using DEM of the given terrain.

Introduction: The ruggedness index is used as a measurement of terrain heterogeneity. The ruggedness index value is calculated for every location, by summarizing the change in elevation. Ruggedness index values have been classified into categories to describe the different types of terrain. These areas of ruggedness are likely found around the peaks of the highest mountains and in areas with large cliffs, where changes in elevation are more distinctive.

Ruggedness Classification	Ruggedness Index Value
Level	0 – 80m
Nearly Level	81 – 116m
Slightly Rugged	117 – 161m
Intermediately Rugged	162 – 239m
Moderately Rugged	240 – 497m
Highly Rugged	498 – 958m
Extremely Rugged	959 – 4397m

Methodology:-

- Click on start button on the desktop.
- Select programs and go to Quantum GIS Lisboa.
- Then go to Quantum GIS Desktop 1.8.
- Click on add raster layer on tool bar and add DEM layer from the saved location.
- Click on raster on menu bar.
- Then go to terrain analysis and select ruggedness.
- Provide output file name and select the destination for saving the file and click on ok.
- Right click on the layer and go to properties.
- Then click on style and select colormap.
- Open colormap, enter the number of entries and click on classify.
- Then click on apply.
- Right click on the layer and click on zoom to layer extent.
- Save the file by clicking on save project.
- Click on file on menu bar and go to new print composer.

- Go to layout and click on add image to add an image, add label to add title, add legend to add a legend, add scale bar to add a scale bar and add North arrow via add image tool.
- Go to file and click on export as pdf for saving the file in pdf format or click on export as image to save the file as image.
- Provide the name for the output file and select the destination for saving the file.

Result:

The ruggedness of the terrain is determined by using Digital Elevation Model for the given terrain.

12. DETERMINATION OF MEAN CENTRE FOR GIVEN TERRAIN

Aim: To determine mean centre by using point features for the given terrain.

Introduction: Mean centre identifies the geographic center or the center of concentration for a set of features. The mean center is a point constructed from the average x and y values for the input feature centroids. The mean center is the average x and y coordinate of all the features in the study area. It's useful for tracking changes in the distribution or for comparing the distributions of different types of features. Potential applications of mean centre include crime analysis, wild life studies and analyzing level of service.

Methodology:-

- Click on start button on the desktop.
- Select programs and go to Quantum GIS Lisboa.
- Then go to Quantum GIS Desktop 1.8.
- Click on add vector layer on tool bar and add point feature layer from the saved location.
- Click on vector on menu bar.
- Then go to analysis tools and select mean co-ordinates.
- Provide output file name and select the destination for saving the file.
- Click on ok.
- Right click on the layer and click on zoom to layer extent.
- Save the file by clicking on save project.
- Click on file on menu bar and go to new print composer.
- Go to layout and click on add image to add an image, add label to add title, add legend to add a legend, add scale bar to add a scale bar and add North arrow via add image tool.
- Go to file and click on export as pdf for saving the file in pdf format or click on export as image to save the file as image.
- Provide the name for the output file and select the destination for saving the file.

Result:

The mean centre is determined by using point features of the given terrain.

<h1 style="text-align:center">13. BUFFER ANALYSIS</h1>

Aim: To execute buffer analysis using point features for the given terrain.

Introduction: Buffer analysis is used for identifying areas surrounding geographic features. The process involves generating a buffer around existing geographic features and then identifying or selecting features based on whether they fall inside or outside the boundary of the buffer. A buffer in GIS is a zone around a map feature measured in units of distance or time. A buffer is useful for proximity analysis. **Proximity Analysis** is an analytical technique used to determine the relationship between a selected point and its neighbors.

Methodology:-

- Click on start button on the desktop.
- Select programs and go to Quantum GIS Lisboa.
- Then go to Quantum GIS Desktop 1.8.
- Click on add raster layer on tool bar and add a modified tif file.
- Open toposheet and digitize it.
- Go to layer on menu bar and click on new.
- Select new shape file layer.
- Then select point type.
- Fill the details in the attribute table like name of the file, type, width to be provided.
- Click on ok.
- Select the destination for saving the file.
- Make right click on the project name and open the attribute table.
- Select toggle editing.
- Then click on add columns.
- Fill the details like name of the column, type, width to be provided.
- Click on ok and close the window.
- Go to toggle editing and click on add features.
- Select point features and provide ID and name.
- Click on vector on menu bar and go to geoprocessing tools.
- Select buffer and provide sufficient buffer distance.

- Provide output file name and select the destination for saving the file and click on ok.
- Right click on the file ad click on zoom to layer extent.
- Save the file by clicking on save project
- Click on file on menu bar and go to new print composer.
- Go to layout and click on add image to add an image, add label to add title, add legend to add a legend, add scale bar to add a scale bar and add North arrow via add image tool.
- Go to file and click on export as pdf for saving the file in pdf format or click on export as image to save the file as image.
- Provide the name for the output file and select the destination for saving the file.

Result:

The buffers and multiring buffers were created using point features of the given area.